Is Biocentrism Dead?

Understanding the Universe and Nature

By: Raj Bogle

Publishers Notes

Disclaimer

This publication is intended to provide helpful and informative material. It is not intended to diagnose, treat, cure, or prevent any health problem or condition, nor is intended to replace the advice of a physician. No action should be taken solely on the contents of this book. Always consult your physician or qualified health-care professional on any matters regarding your health and before adopting any suggestions in this book or drawing inferences from it.

The author and publisher specifically disclaim all responsibility for any liability, loss or risk, personal or otherwise, which is incurred as a consequence, directly or indirectly, from the use or application of any contents of this book.

Any and all product names referenced within this book are the trademarks of their respective owners. None of these owners have sponsored, authorized, endorsed, or approved this book.

Always read all information provided by the manufacturers' product labels before using their products. The author and publisher are not responsible for claims made by manufacturers.

Digital Edition

Manufactured in the United States of America

What You Will Learn In This Book

How This Book Will Help You and Why

Biocentrism is a type of philosophy or ethical way of thinking. To put it simply, biocentrism is a way of thinking that puts inherent value on all living things. Some people may feel that this is a very common viewpoint. In this book we will cover every aspect of Biocentrism to teach readers how this will help them in their everyday lives.

Dive Right into the Book! Or Learn a Bit More About the Author

TABLE OF CONTENTS

DEDICATION

This book is dedicated to my brother that taught me everything I know about biocentrism. Thanks Mike! – This one's for you!

CHAPTER 1- WHAT IS BIOCENTRISM?

A lot of people are curious as to what "Biocentrism" is. Biocentrism is a type of philosophy or ethical way of thinking. To put it simply, biocentrism is a way of thinking that puts inherent value on all living things. Some people may feel that this is a very common viewpoint. Very few people would argue with the value of life. However, biocentrism is the assertion that all living things have value; not just humans. In fact, biocentrism would dictate that the life of a plant or an animal would have just as much value as a human life.

A key component of biocentrism is an attempt to adjust how humans view themselves in relation to the world around them. There's no doubt that there are a lot of humans that choose to hold themselves in higher regard than nature. Deforestation, cruelty to animals, and many capitalistic ventures are evident of this phenomenon. An important tenant of biocentrism is to get humans to view themselves as part of an ecosystem, as opposed to thinking that humans are above or outside of nature. When humans are destroying the ecosystem, we have to think of it as destroying our own home. This is in direct contrast to people who feel that destroying the environment is only a detriment to "lesser" species.

For all intents and purposed, humans are merely animals. Humans are advanced animals, but that doesn't change the fact that a lot of our behaviors can be reduced to a biological and animalistic level. Those who believe in biocentrism deny the idea that human life is, in some way, superior to the lives of other species. They feel that other life-forms have just as much right to do what they want as humans do. A lot of humans view themselves as superior to all other life-forms on the planet. This is due to the fact that humans have a conscience and are capable of higher levels of thinking. Biocentrism asserts that this distinction is entirely irrelevant; other life-forms pursue their own "good" just the same as humans do.

There is a different between biocentrism and other philosophies when it comes to nature. For instance, since environmentalism is such a broad umbrella of ethics and philosophy, one could argue that biocentrism is a subset of environmentalism. However, environmentalists seem to put more focus on the actual environment and how it benefits all life. Biocentrism is a bit more specific in that it asserts that human should want to protect the environment because humans are a part of the environment. With that being said, biocentrists and environmentalists are natural allies. They want the same thing, but the philosophies can differ in a few minor ways. In some ways, environmentalists can be even more damning of humanity. A lot of environmentalists would argue that human life actually has less value than the lives of other species. This is due to the idea that other species do things out of necessity and "natural order." For instance, a wolf doesn't kill a rabbit for fun or for some emotional gain.

A wolf kills a rabbit for food. Some environmentalists believe that humans act in a way that is not in accordance with the natural order of things. Animals don't go extinct due to natural selection anymore. Animals go extinct because humans hunt them down for sport or profit. Since humans are negatively affecting the ecosystem on such a grand scale, some environmentalists would argue that human life is a bad thing for nature. Alternatively, biocentrists believe that human life isn't greater or lesser than the lives of other species. It's just there.

One important aspect of biocentrism is the idea of interdependence. Since humans are just another part of the ecosystem, we have to do our job to ensure that other species are faring well. If they don't, it could have a disastrous effect on the ecosystem and on us. Take the food chain; once one animal or plant is removed from the food chain, it has an effect on the entire chain. Frogs eat flies. Snakes eat frogs. Mongooses eat snakes. If, for whatever reason, the flies go extinct;

the frogs would starve or migrate somewhere else for more food. The snake is without food and many of them will starve. The mongooses then have to find new food or starve also. Biocentrism takes this idea and puts it on a grander scale. Every animals or flower that we destroy is going to have an effect on how we, as humans, live our lives. Take the honey bees, for example. Colonies of honey bees are dying off at a massive rate. Scientists are still trying to figure out what's going on. Why is the honey bee problem a problem? Honey bees provide an essential function of pollinating our crops. An entire 33% of the human diet is entirely dependent on the function that honey bees serve. If the honey bees end up going extinct, there is going to be a massive drop in the amount of crops in the world. This will drive prices up massively and cause a food shortage unlike the world has ever seen. If humans are so superior to other animals and plants, how can we be so harshly affected by their absence? This is the idea that biocentrism is attempting to put forward; that humans are merely part of a greater system and we should do everything we can to ensure that this system keeps going. This isn't just for the benefit of nature, with ourselves outside of nature, but this is directly for our benefit.

Despite its seemingly good intentions, there are many critics of biocentrism. Some critics argue that biocentrism is an anti-human philosophy. However, this critique is in direct contrast with what biocentrism is attempting to assert. It isn't anti-human, but it isn't pro-human. Another critique is that humans are acting naturally. For instance, humans were made the way they are by natural means. This means that the behavior of humans is still in accordance with the natural order of things. This can be considered a glaring contradiction of biocentrism.

Chapter 2- Who Is Dr. Robert Lanza?

Once, the world was flat. There was no gravity. We only had one land mass. These ideas were changed by revolutionary new theories, often not accepted at the time, but later proved to be true. Are we at the dawn of a new change again? Many scientists like Robert Lanza believe this to be true.

Up until the 20th century, scientists believe the earth's continents did not move. During the Greek Empire, people believed the earth was the center of the universe; and for thousands of years, everyone believed diseases like cholera, the Black Death and Chlamydia were caused by a type of 'bad air.' All of these beliefs were eventually disproved, and the new theory to take its place, at the time was always viewed as a "mistaken" theory or idea. These are the changes that have taught and changed us; yet they have always brought great controversy and disbelief.

Today, we are provoking more change than ever before. Scientists and theorists are making discoveries that are turning the world on its head. In fact, some say that the 21st century will be the century of Biology; and it will be this that gives us the basis of the "true" theory of existentialism; this is a complete shift from that of physics used in the last two centuries. It is a simple idea, one concurred with by many scientists but introduced by biologist, Robert Lanza. His idea is that we consider the world biologically centered.

The Theory Of Biocentrism

This is a simple theory, expressed through what Lanza calls Biocentrism; a theory derived from prior experimental findings of quantum theory. Biocentrism is a new paradigm that shows life is not an accident; in fact it is life that creates the universe.

Experienced and great minds like Robert Lanza, suggests that we are the universe. Taking this approach unlocks the blocks scientists have managed to corner themselves into. This theory offers an interesting perspective on life and the world that surrounds it. In essence, Biocentrism completely changes the way that we see life, death even time and space. It demonstrates that life is much more than the activity of carbon, water and a few other elements. This is the theory that may show that life is immortal.

Biocentrism shows a new possibility, a new perspective seen from a scientific point of view. Robert Lanza M.D. is today, considered one of the leading scientists of the world. He is Chief Scientific Officer at Advanced Cell technology, and has made significant research into stem cell research and development. In fact Lanza's, work has had a significant impact on our understanding of nuclear transfer of stem cell biology.

His recent book, Biocentrism, centers on how life and consciousness are the true nature of the universe. The book takes a scholarly and scientific stance, but also brings philosophy to the table. Robert Lanza deemed one of the most brilliant minds of our time, has worked on this new Theory of the Universe, based on the knowledge we have acquired in the last few hundred years. In this theory he suggests that it is our perspective on our biological limitations that have limited our understanding of our existence.

Consciousness Creates Reality

This new idea that consciousness creates reality is not just an idea that swims out in philosophical musings. It is based on biological findings and on quantum support. It adheres to the fundamentals that biology and neuroscience tell us about the structure of our human form. The theory works like this:

Just as we know the sun doesn't move about us, but we move around the sun, that we are the one doing the activity, so are we the beings that give meaning to our reality, suggests Ronald Green, Director of the Ethics Institute, at Dartmouth College.

The Essence Of The Biocentrism Theory

Lanza's theory is consistent with some of the world's most ancient traditions; those which suggest that it is consciousness that governs, conceives and becomes physical. This book places life as the center of the universe and not as an accidental byproduct.

This book generates controversy because of its argument that consciousness is central to creating the universe. Lanza believes that science is trying to understand the world from a completely wrong perspective when they treat space and time as a physical thing. Other physicists also believe Lanza's point of view is in perspective and that it aligns with quantum mechanics.

What makes this a must read is that Lanza is the first scientist to stand up and say it out loud. Physicists who have come to similar findings only whisper of this new theory, believing in its possibility, but not considering it to be politically correct.

Is Biocentrism the Answer?

It may still be too early to say. But what we do know is that this scientific study has even the brightest minds in the scientific world wondering. No one denies that the current theories on our existence are muddied. There are unexplained paradigms throughout. And it is through studies like Biocentrism that a new view point may be the answer to the intertwined invisible fabric of the physical world and the conscious one.

Bottom Line

Some scientists insist that a new Theory and answer to everything is just around the corner. Others may whisper of it, but are reticent to stand up and suggest that everything we have based our beliefs on is wrong. Even so, most say that any day now we will fully understand the cosmos, we will understand who we are and know everything about the world that surrounds us. This indeed is a new revolutionary idea and Robert Lanza is at the very foundation of this life-changing theory.

Scientists now believe that consciousness could be at the very core of our existence. It is not a small item as we once thought. It is the awareness and perception that has arisen from mere molecules which have joined together. This is the very explanation that which defines what makes us different from one another, yet so similar.

CHAPTER 3- THE SEVEN PRINCIPLES OF BIOCENTRISM

Biocentrism is a theory proposed in 2007 by Robert Lanza, a medical doctor from the US. It is a controversial view that casts doubt on the dependence of the scientific method of objective observation and measurement, as the sole means of determining reality. In the book that he authored along with Bob Berman entitled Biocentrism: How Life and Consciousness Are the Keys to Understanding the True Nature of the Universe, Lanza states that life creates the universe, not the other way around. It is explained that biology is the central driving force among the sciences and is the means of understanding the other branches of science such as Chemistry and Physics, expanding them on a deeper level.

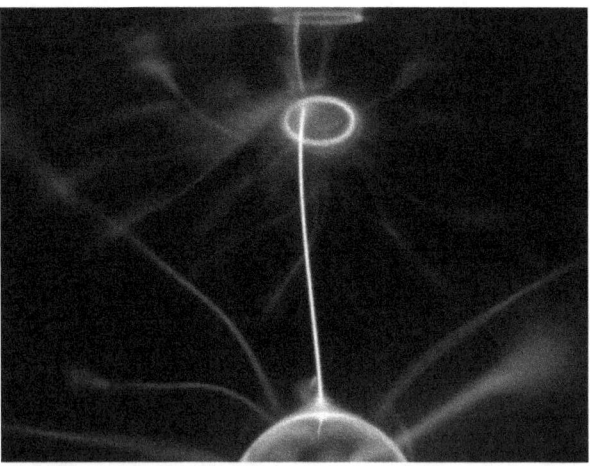

A Background On Biocentrism

Biocentrism states that biology and life itself are at the center of being, reality, and the cosmos. It shows how Biology can build upon Quantum Physics. Lanza was inspired by the 20th century physicist John Archibald Wheeler as he formulated his hypothesis that "life creates the universe". He has put a mystical spin on Wheeler's work and developed it to effectively converge with current theories and research. He utilizes relational quantum mechanics as a basis for his theory.

Lanza makes a case that the theories that have been developed that try to explain how the world physically runs, fall short due to the fact that they do not address deeper understandings of consciousness. Currently the various sciences have their own areas, but biocentrism suggests that scientists should all turn to Biology before approaching the other sciences. Lanza makes a case that Biocentrism possesses the potential to unite all the various branches of physics, something not even Einstein could successfully prove in his unified field experiments from eighty years ago. According to Lanza, biocentrism offers insight into several things that have perplexed scientists for years.

The First Principle Of Biocentrism

The first principle states that what we observe is dependent on the observer. While we look at something as being reality, it is actually a process of our consciousness. While we have traditionally assumed an external, observer-independent world, the revelations provided by quantum mechanics have given rise to challenges in this; a "superorganism" that interfaces with the universe, setting a course of universal events into motion. Quantum Mechanics refers to the "delayed choice experiment" in which a decision made by an experimenter seems to change the way a particle behaved at an earlier point in time. We are not subject to an absolute course of predetermined events. It seems as though the observer does play a role. Biocentricity explains that inanimate objects that are built by conscious humans may also function as observers, detecting information by seeking it out. Their design often imitates the living organism.

The Second And Third Principles Of Biocentrism

The second and third principles state that our external and internal perceptions are interconnected and the behavior of particles relates to the presence of the observer. Theories of biocentricity suggest that the universe is as complex as it is, due to the diversity of its inhabitants, the biological organisms, who are very complex in nature. Although as humans we have developed technology to examine the universe in precision, we have not achieved an advanced form of consciousness.

The Fourth Principle of Biocentrism

The fourth principle says that consciousness must exist. Without it, matter stays in an undermined state of probability. A simplified and stronger model of the emergence of the universe emerges when we are willing to call in a new perspective. The scientific knowledge gained from this perspective that is consistent with experiments can lead to new perspectives becoming physical reality. The biocentric universe orchestrates the physical relation between biology and the physical world. This relationship is what determines the principles and laws upon which the universe operates.

The Fifth Principle Of Biocentrism

The fifth principle points out how the structure of the universe itself has laws, forces, and constants that appear to be perfectly adjusted towards life. The biocentric universe theory is an alternative to the typical ways that scientists have tried to explain a view of the world and our place within it. Biocentrism challenges the standard thought that the universe has been formed by atoms and particles coming together to create life forms, pointing out that this theory is lacking in evidence. It suggests that the universe probably started off very simple and has gradually become more and more complex, fine-tuning itself to accommodate matter and life. Each observation is consistent with another, creating harmony. Two different things may in fact both be true. Within the western scientific tradition, generally people have subscribed to a metaphysical system that is absolutist to describe physical reality. This is why some physicists have worked so hard to square quantum experimental findings with this absolute paradigm.

The Sixth And Seventh Principles

The sixth and seventh principles state that time and space are not objects and things, but tools that we use in our understanding. Lanza states that we are like turtles in their shells carrying around space and time with us wherever we go. He points to the works of famous philosophers such as Descartes, Kant, Schopenhauer and Bergson to show that an emphasis on consciousness is of primary importance. Our individual notions of space are a result of personal observations and the way we relate one thought or event to another. These observations are ordered in our mind in a way that changes our perception. This is how we conceive of time.

In short, Biocentrism suggests that life is pivotal in our understanding of the universe and it is not an accidental or random byproduct of physics. Biology includes everything and there is no real existence outside of it. Manifested features within the universe come about as a response to consciousness. Biocentrism offers a shift in the interpretation of scientific knowledge that has the potential to lead to new understandings of philosophical and physical conundrums. It potentially resolves the many paradoxes in science.

CHAPTER 4- BIOCENTRISM'S TAKE ON THE COSMOS

From time to time, a new and radical idea will come along to challenge everything we thought we knew about the universe. One of the most radical changes to our understanding of the universe came in 1926 when Erwin Schrodinger crafted the Schrodinger equation and the field of quantum mechanics grew up around it. Quantum mechanics took the idea of certainty and threw it out the window. Nature could no longer be described with certainty; instead quantum mechanics teaches that nature is just a series of probabilities; a particle is probably located over here, but there's also a chance it'll be over there.

Perhaps one of the most interesting ideas to come out of quantum mechanics is the idea that nature can be affected by observation.

Quantum mechanics shows us that elementary objects, like photons, are both particles and waves at the same time. In the Double Slit experiment photons are shot at a piece of photographic paper behind two slits. If a photon is a particle, it will go through either the left slit or the right slit and leave a single point of impact of the paper; if it's a wave it will travel through both slits and create an interference pattern. When the experiment is run, the photographic paper shows both point impacts from particles and interference patterns from waves; photons are both a particle and a wave at the same time. When you add a human observer to the mix, the results are even more interesting. If a human observes the experiment then the photon will behave as either a wave or a particle, but not both simultaneously.

The implications of this are staggering. Life, just by the simple act of observation, can change nature. That photon is forced to change into either a particle or a wave by human observation. This is a world shattering idea, one that changes everything we think we know about nature. Most people have probably heard the story of Schrodinger's Cat, a thought experiment designed to show the implications of the act of observation. In this experiment, a cat is placed inside a sealed box and cannot be observed. The box is attached to a device that has exactly a 50/50 chance of killing the cat; the question is: is this cat alive or dead? Our normal understanding of the universe would tell us that the cat is either alive, or it is dead, we just don't know until we open the box to check. Quantum mechanics, on the other hand, shows us that this is not the case. That cat is BOTH alive and dead at the same time until the box is opened, and the cat is observed. Until something is observed it exists in every possible state simultaneously; it is only the act of observation that forces something into one particular state (either alive or dead in the case of the poor cat.)

This is the power of observation; this is the power of life. Life literally shapes the world around it. It is our perception of the universe that gives reality form. Physics is an amazing science; it has given us tremendous insight into the universe. But physics, by itself, is incomplete. Why does observation force a photon to be either a wave or a particle; why are all the forces of the universe balanced so perfectly for life? On questions like these physics is silent.

Why is the cosmos so perfectly tuned to allow for life? If gravity had been just the tiniest bit stronger, or weaker, stars could not have formed, planetary systems could not have formed; life could not exist. A slight increase in the strong nuclear force would have automatically fused all hydrogen into helium; water could not exist. Without water life as we know it could not have existed. So why are all the forces of nature so perfectly balanced to allow for life? The best answer mainstream science can give is called the Anthropic Principle: the idea that the universe must be tuned to allow for life because if it weren't we would not be here to ask why it was so well tuned. That is an unsatisfying answer at best.

There is a better answer. The universe is so perfectly balanced for life because life made it so. Life made the universe, not the other way around. This is the idea of biocentrism; it is the idea that it is biology, life, which is the central driver of the universe. It is the idea that the best theories in physics and cosmology are incomplete, and will always remain incomplete, because they do not take the role of consciousness into account. Physicists have long struggled to produce a "theory of everything," one single theory that could explain everything about the cosmos. They have failed, and they will continue to fail until they come to realize that a theory of everything is impossible unless they incorporate the power of life and consciousness.

Biocentrism explains much of the things that so puzzle scientists. The double slit experiment, the perfect balance of forces in the universe, the questions of where the universe came from, all suddenly make sense when you realize that life, and consciousness, come first. Photons are forced to be either waves or particles when observed because life shapes reality. The universe is perfectly balanced to allow for life because life made it so. We created the universe. The last is probably the hardest to understand, how could life create the universe when the universe has existed for billions of years before the appearance of life? The answer lies in physics. Quantum mechanics demonstrates that, until observed, a system will exist in all possible states simultaneously. This is the early, pre-life, universe. Not the cosmos as we currently see it, but a sea of different possibilities all overlapping and existing at once. Universes that are completely hostile to life, universes where stars can never be born, universes where gravity is so weak that nothing larger than a grain of dust could ever develop, and in all the myriad possibilities existed a universe where life was possible. When life arose and looked up at the cosmos that simple act collapsed all of the countless possibilities down into one: the universe as we know it.

The cosmos is an extension of our consciousness; it is the idea that consciousness shapes reality. Without consciousness matter exists as an undetermined state of probability, just as Schrodinger's cat is both alive and dead until observed. Space-time is not an object separate from life; it is shaped and created by life. Biocentrism shows us that there is no independent external universe outside of life. Life is central to reality and the cosmos. This is the beauty of biocentrism.

CHAPTER 5- THE BIOCENTRIC VIEW OF CONSCIOUSNESS

Biocentrism is a relatively new scientific theory which proposes that reality is merely a translation of energy reactions by our conscious perception. Without consciousness we would have no awareness of principles like space and time which govern our scientific view of the cosmos. As I said previously, Robert Lanza, an American doctor who first proposed his ideas of a biocentric universe in 2007, tied consciousness to the creation of the universe itself. In this way of thinking consciousness and life actually created the universe, not the other way around. Dr. Lanza stated that modern science needed to realign its concepts of the universe with biological principles rather than using physics as its fundamental laws. He proposed that the origin and laws of the universe only made sense if viewed from a conscious perspective. In a biocentric universe life and biology are central to our understanding of being, reality, and the cosmos.

As Mentioned in Chapter 3, Dr. Lanza wrote a book, entitled "Biocentrism: How Life and Consciousness Are the Keys to Understanding the True Nature of the Universe," in which he sets out seven principles that outline this biocentric universal view. These principles explain that what we experience as reality is really only a reflection of our brain's perception of signals and reactions going on in the universe around us. Our scientific principles of space, time, light, and sound are all reliant on our brain's conscious perceptions of the data around us. Space and time are not constants of the universe because they change relative to speed. This idea that the principles of physics are really just how our brains create connections In the things we are observing fundamentally changes the way we define the universe. It means that there really is no independent external universe outside of biological existence. Objects, our physical world, and all that we base our laws of the universe on are internal perceptions. In other words, the brain is like a super computer of consciousness which analyzes the data around itself to create reality and without it the universe, in essence, would not exist. Many people relate this idea to the "Matrix Principle," where we are simply consciousness floating in a stream of data that we constantly process into what we understand as reality.

Dr. Lanza hypothesized that the universe itself was formed by consciousness with the specific objective of supporting life. He argues that the concept of the universe as random chance is entirely improbable, and he cites over 200 physical parameters within the universe that are so exact they could not have been probable without factoring in the existence of life and consciousness. Without consciousness a universe can only exist in an endless undetermined state of probability. This is known as part of the observer effect, where without a conscious observer things exist in underdetermined waves of probability. Taking Dr. Lanza's idea of consciousness as the root of reality, the observer is actually defining the reality of subatomic particles. Without an observer what we define as particles exist simultaneously in time and space meaning they have an undetermined future of probability. Consciousness creates our reality by collapsing this probability cloud of the subatomic particles. This theory concludes that the universe was created by the introduction of consciousness to particles which were existing in an undetermined state of probability. Along these lines, through the introduction of consciousness to the universe particles were set in motion down a specific path that would lead to the creation of life in order to support that initial consciousness.

That is quite a lot to swallow for some modern scientists, most especially physicists. Dr. Lanza's theories of biocentrism and a conscious universe have been criticized by some as a sort of wild philosophical theory rather than based in scientific principle. Some scientists seem to agree that many of Lanza's theories currently don't provide testable predictions and therefore cannot be taken seriously.

However, there are many scientists who have taken Dr. Lanza's ideas much more seriously, even expounding upon his theories. Some scientists have even said that the definition of a biocentric universe meets up with many wisdom traditions around the world. No matter what it can be said that biocentrism answers some interesting questions in scientific theory, while at the same time creating even more interesting queries.

If this theory is taken seriously, how would one test for its validity? What could it mean for the world of physics? If consciousness creates the physical world, then would it be possible to change our perception and therefore change what we know as the laws of physics? Another interesting line of questioning that has resulted from this theory is the idea of collective consciousness. Many believe that this theory of consciousness would explain things like telepathy, psychokinesis, and the placebo effect. If consciousness is interacting with other consciousness, then it might follow that we could tap into this pool of universal consciousness. It is these types of extraordinary theories that cause the scientific community to cast a shadow of doubt of Lanza's scientific theories. Currently Dr. Lanza is continuing his research on consciousness, and many hope to see interesting future experiments as Lanza tries to prove this biocentric theory.

Whether or not all of Lanza's ideas can be proven, he casts a doubt on the idea that all of science's questions can be answered via the use of objective measurement and observation. It seems Einstein's questions of time and space have really taken modern scientists to the edge of measurement and beyond into a universe of uncertainty. Robert Lanza proposes some interesting answers to these age old questions, but he uncovers even more baffling lines of query and experimentation. It is unclear how Dr. Lanza intends to definitively prove his biocentric theory of consciousness, but he has ideas of a quantum experiment that just might be able to prove (or disprove) his overall theory. Either way, in the next wave of scientific discovery we might find ourselves just a little bit closer to answering some of the universe's biggest questions.

CHAPTER 6- DEATH AND ETERNITY FROM A BIOCENTRIC PERSPECTIVE

Death and eternity are concepts that pry at the curiosity of all living beings. These ideas provoke questions of what happens to us once the present deposits the current moments into the past or once the future intersects the here and now. Are concepts, such as death and eternity real ideas, or simply convenient patterns of thinking that helps a member of a species recognize their current place in the stream of events that passes through our conscious minds? In a Biocentric model of the universe, such concepts, whether they truly exist or not, cannot be discriminatory. If death occurs, it is a state intended for all species. If time exists, it is felt by all who experience the passage of events under some grand order of operation as played out within the mind of the individual member of a species. And if these concepts are manufactured by convenience, it is a convenience in which we all inherently indulge by virtue of the way an individual's mind processes observable phenomena.

Death Eradicated By The Multiverse

According to Robert Lanza, life and its associated consciousness cannot be snuffed out. Much like a cartoon, where Bugs Bunny is seen being annihilated in one frame and reanimated in another, the preservation of life and consciousness in a biocentric model is preserved inside the context of a principle of conservation of existence. The implication arises from the broad perspective of the universe as a multiverse containing an infinite number of universes.

Inside this multiverse, anything that is possible to occur, does occur. At least, it occurs in some alternate universe, if not in the universe in which we currently reside or in which our mind currently takes conscious focus. Hence, if a person seems to die in one universe, their existence and consciousness is preserved in the other universes which contain the alternative possibilities not contained in our present universe. In this sense, in the context of the multiverse, the individual member of any species is essentially granted a potentially scientific basis for the doctrine of the immortality of the soul being extended equally to all individuals of all species. However, this idea is not new or unique to a modern biocentric community, since even in early Jewish and Christian theology, the entity Lucifer is reported as claiming:

- Gen 3:4 And the serpent said unto the woman, Ye shall not surely die:
- Gen 3:5 For God doth know that in the day ye eat thereof, then your eyes shall be opened, and ye shall be as gods, knowing good and evil.

Augmenting this idea by Lanza's multiverse perspective implies that even if the Christian God is powerful in this universe, the multiverse preserves the possibility that in other universes God is not powerful enough to revoke the existence of others, such as Lucifer himself. Hence, a biocentric premise, contained inside the context of a multiverse that enumerates all alternative possibilities, inherently aims to debunk the claim that any God of any religion is truly supreme in power, hence establishing Robert Lanza's preservation of existence of an individual across the multiverse as a more multiversal conservation of existence principle.

Linear Breaks In Existence

Imagine watching a favorite television series. It could be the many seasons of Gun Smoke or some other epic saga, for example. As the series winds down to the last season, as the last episode scrolls across the screen, the lifespan of these characters essentially comes to an abrupt end. This end is much like the experience of death in the biocentric multiverse. The energy that your mind harbors, the feeling that encapsulates who an individual believes they are in this life reaches an end, much like a low wattage light bulb burns out after some period of time. However, since the energy is conserved, life is never truly extinguished out of existence. Rather, this end is only realized in this particular single universe of the broader multiverse. Since all possible universes exist at the same time in the multiverse, death at most only implies a transition of consciousness to another place in the multiverse. It's much like starting a new television series. The people are not the same. The scenery, relationships, circumstances, and numerous other details are different as well. But, the observer who experiences this transition is the person who existed before the last break in what our brain might cause us to imagine is a relatively linear progression of events. Such ideas as these are now being entertained and taken more seriously by a number of physicists who believe in the existence of the multiverse at a minimum, and from there accept some of Lanza's characterizations of how existence plays out for an individual across the multiverse in general.

Eternal Assurance In Biocentrism

If the energy, essence, or soul of the individual is itself essentially immortal, being that energy and existence are conserved in the multiverse, this would tend to suggest that eternal immortality is guaranteed regardless of an individual's moral state. In many ways, such a model is the antithesis of other religious models which grant eternal life, rewards, and rites of passage based on some deity's evaluation of an individual's good or bad behavior. The idea is that nature, unlike an intelligent god or goddess, may not place any specific value on behavior at all. Hence, the multiverse exists not to test for behavioral progress, but instead, simply exists to process the phases, breaks, and transitions in existence from one portion of the multiverse to the next. In this respect, some might conclude that a biocentric model implies that nature is by default a system where all individuals are equally granted the right to eternally exist and therefore there is no difference in the value of one individual over another. In one life nature places a person in a prosperous position in society, but in the next life they may have a more humble existence. Not because they necessarily deserve one position or the other, but because the diversity they experience from one life phase to the next is merely the inherent spice of life, which comes as a result of being part of such an eternal existence.

CHAPTER 7- BIOCENTRISM- ARTIFICIAL LIFE AND THE FUTURE

As you now know from previous chapters, the idea of a Biocentric Universe was first suggested by the Robert Lanza, a doctor from the United States. Lanza's view sees biology at the very center of the universe. Biology then acts as the main force that drives, pushes, and regulates every other force that we experience on earth. Lanza proposed that when we delve further into other sciences, we are making arbitrary distinctions between fields, and really we are just learning more about biology. The main idea differs so radically from more streamlined and established understandings of the universe; that life is what has created the world around us, and not the other way around.

Lanza's unique background is surely what led him to the Biocentric model of the universe. As a doctor of regenerative medicine and biology, Lanza played a major role as a member of the stem cell research lab. There he was a part of several projects that worked on cloning embryos. It is this rich background in biology that opened his mind to the notion of Biocentrism. Lanza's ideas first appeared in the popular and influential publication, The American Scholar, where they were met with some success. Enough success apparently that Lanza later published a book that served as an extension and clarification of the ideas that had appeared in The American Scholar years earlier. These ideas have sparked new insights into the way people across all fields are looking at the phenomenon of existence. Lanza's views have begun a brand new manner to see not only our present situation with fresh eyes, but our futures begin to take form in an entirely new manner.

One important bit of the future that Biocentrism radically changes is the idea of death and what happens to us after we die. Lanza says that we simply have a belief in death, yet this mere belief doesn't translate into the pure reality that we perceive. How can this be? Well, it starts within the body. We see the things we see, we do the things we do, because we are programmed to see them and do them in the manner that we do. It all comes down to our genetics; the structure of the chemicals that drive us, the receptors that tell us what it is we are thinking. We live entirely within our own minds and bodies. So, what does this tell us about death?

Let us consider for a moment the Heisenberg Uncertainty Principle. It states that we, because we can't measure both a particle's momentum and location simultaneously, there's no "actual" world out there that we are trying to measure. Things are partially a product of us; our minds. The application to death is then that death cannot be a state in which there is no time and no space. This is because the existence of time and space as we currently conceive of them are simply products of our current perception; a perception which we now know to be fatally flawed. It is here that Lanza and similar scholar suggest that could be an existence outside of time, where we are somehow free of our limitation of the insistence of time and space as the monarchs of our world.

Another experiment that Biocentrism relies upon is an experiment that was done on pairs of light photons. These photons seemed to have the strange ability to predict what the other partner in the pair would do, and then mimic that action, before the first action had even occurred. This suggests not that the photons have psychic powers, but rather that the photons are in another state of existence, an existence outside of time in which they are acting simultaneously. It is us the observers who perceive the actions as occurring at different times.

These experiments open our minds to the possibility of an existence without the current limitations that we experience, and this is likely our future after the grave. Consider the multiple universes theory. This states that everything that we observe that could have gone differently in our world, did in fact, occur the other way in a different universe. This complex of endless worlds is known and the multiverse, and it is within this multiverse that all possibilities are exhausted. Our matter, our consciousness in existence in another multiverse is likely our future if we subscribe to the philosophy of Biocentrism.

Additionally, Biocentrism adds a new facet to the many opinions about artificial life. A major component of Biocentrism is that all biological beings have an inherent morality and reverence to their life. Elements of Biocentrism maintain that artificial life has the same inherent morality, and also suggests that scientist and those who help create artificial life have a commitment to respect and take responsibility for the life that they create. However, even in regard to religious concerns, Biocentrism upholds the mindful creation of artificial intelligence.

Moreover, the future may even be more a part of our current lives than we could have ever imagined according to some of the implications of Biocentrism. In fact, studies have suggested that the future may actually be impacting life as it exists now. The same experiment with the twin light photons can also be interpreted not only as a possibility for a future state of existence, but can be viewed as an application of our current circumstances. The experiment can help us conclude that our future is impacting us now. It implies that if these light photons can be influenced by the future, then surely our surroundings can be similarly manipulated. We may be stuck in a linear view of time within our current existence, but this doesn't mean that we don't experience the effects of the future upon our present without realizing it. Biocentrists appear to agree with the genius of the philosopher Spinoza in his ideas about all time: past, present, and future, being inextricably intertwined. Furthermore,

Spinoza noted that our recognition of this phenomenon can actually transcend the barriers of space and time.

About The Author

Raj Bogle was always intrigued by the sciences and as a result of this he ended up majoring in the sciences so that he could have a better understanding of life. When he was doing research he became increasingly interested in the concept of biocentrism that was coined by Dr.Robert Lanza which purports that life creates the universe. It was a new way to look at things and definitely caused even more questions to be asked about the universe.

Raj has taken what he has learned over the years and also taken Dr. Lanza's theories and has come up with his own view of the topic.